The Information Technology Career

THOMAS S. IRELAND

DEDICATION

To Information Technology Leaders and
Director Contributors Everywhere.

OTHER BOOKS BY TOM IRELAND

Information Technology Leadership Books:

How To Write A Great Information Technology Strategic Plan – And Thrill Your CEO

How To Create A Great Information Technology Vision – And Thrill Your CEO

Information Technology Leadership Excellence

Rules for Information Technology Leaders

By Tom Ireland and Sandra Mikesell

Program Management: The Path to Leadership

And Tom Ireland's Science Fiction Novel

Quantum Soul

Contents

Introduction

OK. You've decided to make information technology a career. Or, perhaps you just ended up in IT by accident and see it as a fun and interesting thing to do. Well, as a forty-five year veteran of the IT wars I can guarantee you that you are right. It will be interesting! And you can take interesting to mean almost anything you want it to mean.

Now that you have entered IT, or are on the verge of it, I hope you will find out that you love it. Because you will need to as it tears away at your sanity over the decades. As much as you might love your time off you will need to be so enamored with information technology that you will be as excited to return to the job at the beginning of your work week as you were tired at the end of the previous week.

Make sure your significant other understands that you will never truly be off duty. Computing, voice communication, and network systems break because they are made by humans like you. And these other humans aren't any more perfect than you are.

Naturally, any IT system will break at the worst possible time; during a romantic dinner, a ball game, or perhaps that vacation that you are supposed to be on. Or even while doing something "very special". That's just the way it is. You will learn that the famous guy named Murphy that all techies refer to was actually an optimist. It is far worse than anything he describes.

I was in the middle of hanging wallpaper one weekend when we had a system failure at work. I was clearly rationalizing when I figured it was something minor and it could wait. So much for rationalizing! When I got to work on Monday it was a bit of a mess which we cleaned up after a few hours and a lot of harassment from sales people trying to calm down our customers.

That was the easy part. When we had everything squared away the VP's secretary invited me up to his office. Yeah, it wasn't pleasant. I assumed I was going to get fired. That's when it got worse. My VP concluded our 'discussion' by saying, "I'm not going to fire you. I wouldn't do you that favor". He understood how thankless and punishing this amazingly exciting career field is and that I still love decades later.

By the way, you won't be very far down this career path when you will be on a first name basis with Murphy. Heck, you will even be having drinks with this virtual partner at your favorite watering hole after only a couple of years. You may start carrying a small piece of wood in your pocket to knock on for luck a couple of times a day. Oh, I should mention that all IT people eventually become horribly superstitious.

So, if you are still interested let's get into the meat of how to make a career out of an information technology job. Let's also discuss what it takes to become an IT leader.

The Choices

The information technology career field is amazingly diverse. For example, I have held IT jobs in the military, commercial enterprise and local government. I have been an outside plant engineer, quality assurance expert, program manager, developer, installer, operations specialist, manager and executive. I've led organizations of up to four hundred twenty-five people scattered across North America and small operations of five amazing professionals. These jobs have taken me from the Middle East and Europe, to north of the Arctic Circle, to isolated mountain tops as well as the traditional office and operations center. It has been a phenomenal trip!

You can do anything you want in information technology for any kind of company because IT permeates everything in our lives. Just think about that for a few minutes and you will understand that this statement is true.

Information technology will thrill you and you may hate it in the same day or even the same hour. If you are truly serious about your job you will sometimes wonder if you are using your technical knowledge to do things to people or for people. Just read or watch the news and you will understand that this statement is true also.

If, as a responsible person, you think about it a little you might be terrified by the extent of your responsibility. If you are really strange you just might be awestruck by your power. Information technologists build, operate and manage the systems that make much of our national and international infrastructure possible. Errors might cause phones and computers to stop which can bring an enterprise or the entire national telecommunication infrastructure to grind to a halt. Or, you might cause trains or planes to collide with massive loss of life.

So, it is a very serious career field with the potential for huge responsibility for which you will receive very little thanks, but periodic scathing criticism. It is a career field in which one aw shucks wipes out a thousand attaboys.

You will learn something new every day! It is the most rapidly evolving of career fields. Think about it. How quickly does your computer or your smart phone become obsolete? It is a career field in which you can legally be the electronic pusher of new systems and capabilities to a whole planet full of technology junkies. And there is no end in sight to this opportunity so your career is just about guaranteed if you can keep up the pace.

If you can't keep up then you can become an archeologist or anthropologist. The bones and building blocks of ancient civilizations will not move much and will patiently wait for you to discover them.

Two Career Paths

In broad terms, and just as is true with almost any career field, you can follow two general career paths in information technology. You can be a directly contributing technologist, or you can move into management.

Either way it is an opportunity to play with some great and expensive toys (tools) you might never even get a chance to explore if you weren't in IT. Some of this stuff costs millions of dollars and they do really cool things! It is Disneyland every day. And you meet very interesting people.

For example, for a few years I had the opportunity to go to Bell and AT&T Labs to meet true genius scientists and developers. They were overwhelmingly excited to show off their inventions in the hopes that we would take them into the mainstream. It is an amazing privilege to be briefed by a world class brain who hasn't washed their hair in days and wears combat boots without socks under an impossibly wrinkled print dress. It only takes two seconds to get past the visual so you can soak up what the genius has to offer. One briefer walked to our conference room backwards so the scene in front of them wouldn't distract their attention from the mind problem they were working on. Now, how many career fields can you work in that give you that kind of experience?

Whether you are a direct contributor or manager you will have the chance to periodically meet and brief the most senior management. After all, IT permeates every level and each nook and cranny of enterprises of all kinds. This may be true even if your assignment takes you to a third world country. Since you are the one who is commanding the various electrons and photons to go to those locations you will have plenty of opportunity to follow them whether they have done good things or bad. Naturally, you will be appropriately rewarded or punished. Just remember that "No good deed goes unpunished". Sometimes I think that phrase was created by an IT professional.

The Direct Contributor Path

You may be a person who has more success at, and is more comfortable with, talking to electrons or photons than people. If so, you may be better off sticking to being a direct contributor technologist for a few years, or even throughout your career. If you enjoy the one-on-one, direct interaction with the tools (toys?) of the trade then this may be the path for you. And it is a pretty darn good path. You will get the chance to become expert at very complex systems that are the underpinnings, the very foundation, of our modern world. You will be one of the gurus who truly knows things that most others only think they know. (More about that later.) But, be warned. Once you become regarded as a guru it will be assumed that you know everything about information technology.

Let's make a comparison. Physicians always specialize. Even general practitioners specialize at only having general knowledge and the public understands that. Information technologists, on the other hand, might specialize, but are often expected to be experts at the entire broad range of IT. Any reasonable person would admit that this is impossible. But it is clear that customers of IT people are not reasonable. And they are that way for an unreasonable reason.

After all, every family now has at least one person who knows their way around a personal computer, a smart phone and a wireless LAN. That little bit of knowledge, with extensive technical or engineering education, makes them an expert in the eyes of everyone they know. So, given the amount of education and training you have you must truly be the ultimate expert at everything IT. And when you are asked a question, if you don't know the answer, you are suddenly regarded as being of questionable value. Would a brain surgeon be considered useless simply because she didn't know how to do a knee replacement?

That is clearly unfair to IT people. So what? Get over it! Life isn't fair. If you don't like this situation then you may not belong in IT. On the other hand, if you are that rare masochist who loves this kind of punishment at weird hours then we have a real deal for you! You'll have more fun than a water drop on a hot skillet!

To Specialize or Not to Specialize?
Notwithstanding the previous discussion, you do have the option to specialize, or generalize, in information technology. And there is also a middle ground.

A description of a middle ground might be the computer programmer. A programmer might have a great talent and knowledge for talking to a computer but get lost when it comes to integrated wide area network design. Those are two widely divergent disciplines with greatly different ways of approaching systems. So, the programmer might be considered to be specializing in computer programming, know quite a few programming languages and be very adept at them. This programmer could be considered to specialize but also generalize within programming. Oops! Maybe not! Go to Wikipedia and search on "list of programming languages". It goes on for pages and pages! Is this programmer a generalist or a specialist within computer programming?

One thing about information technology is that you are faced with endless opportunities. On a more challenging day you might consider them to be insurmountable opportunities. If you get bored in IT then it is your own fault.

The Opportunity to be Obsolete

Information technology may be the most rapidly changing of career fields. Every day there is the opportunity to learn and apply something new. Developers and vendors are always competing for the consumer's dollar. As an enterprise technologist you may have a lot of dollars available to you for experimentation. If you are one of these lucky individuals then each experiment will increase your knowledge and your value to the enterprise. You will become a vast treasure trove of information that will grow over time. If you are expert at systems integration you may have the exciting opportunity to put this knowledge together in innovative ways to improve profitability or even the human condition. This may be as exciting and rewarding as any person can hope for.

Along with this comes an interesting situation. Remember that IT is one of the most rapidly evolving of any career field. You get to learn new things at the speed of the Starship Enterprise. Sometimes your brain will hurt from trying to pour so much into it so quickly. And virtually none of it is simple stuff. You will accumulate so much information so quickly that the pace of change will eventually make you an expert at everything obsolete because you will learn it so well you can't get rid of it. Sometimes it would be nice to be able to erase parts of your gray matter storage system. But, I guess that could be said about life in general.

However, this knowledge of obsolete systems really is useful. You will eventually learn that computer busses have a lot in common with network backbones. Program languages have structures that are common across languages over decades. All the knowledge that you gain will help you learn the next piece of information even more quickly. All previous technology is a platform for the next discovery. You can't help but be excited by this amazing challenge that will reconfirm your initial thought that IT would be interesting.

All of this technology is created by humans. Therefore it fails. Not only does it fail but it fails in incredibly complex ways. You see, it was probably never properly developed in the first place. Something as complex as IT simply can't even be close to perfect. The newest systems have become so complex, and so software based, that they might have millions of lines of computer code. In fact, an Android smart phone will have about twelve million lines of code. There will be problems and there will be errors. Some errors might be hidden behind other errors. You can be certain that no program this large and complex will be perfect. Correcting these errors is like walking through an endless series of rooms. As you fix the problems in one room you will notice a door at the other end of the room. That will lead to more problems. Of course, if you are an optimist you might consider these as opportunities. The rooms never end.

If you like these kinds of challenges then information technology is definitely the right place for you. It is a career field in which the product will never be perfect. You will have to settle for making it good enough to get the job done.

Now, that can be a scary prospect. Just think of how many lines of code are in the air traffic control system, the GPS network, a space station, the water supply control system for a major metropolitan area, the safety devices in automobiles or the national rail system. Now we are into trillions of lines of code overall which we have to recognize are not perfectly written. They can't be because that complexity is beyond anyone's ability to make absolutely correct. Instead, they just have to be good enough to not kill anyone.

Getting to the point of being good enough means going through alpha and beta testing and extensive field testing in a deliberate effort to break the new or modified system under the worst possible conditions. And you will do this while firm in the knowledge that your system will still have flaws.

If you want to be an information technologist this could be part of your reality.

Yes, this discussion may be a little extreme in the potential consequences but it is still a legitimate discussion that illustrates the complexities and potential consequences of this important and demanding career field.

A more common situation is faced by the IT expert who works for a private enterprise in competition with other corporations to attract customers to their product. For the technologist, this means providing electronic tools that help the enterprise be more competitive by improving efficiency, reducing cost or creating a better, higher quality product. Remember that IT is a critical component in the infrastructure of almost all enterprises worldwide. The IT piece of the infrastructure can mean the difference between success and failure of that enterprise.

The Leadership Career Path

At some point in your IT career you may decide that your professional goal is to move into management. Remember when you make this decision that you are taking on the task of leading geeks and nerds. Since you may be one yourself you will probably recognize that this is a challenge involving leading well-disciplined, dedicated people working for the betterment of the "collective" but with strong-willed ideas about how best to do things. For a self-confident leader with a clear objective and who values the opinions of others this is a great opportunity. For the leader who cannot clearly state a strategic objective in both technical and business terms this is suicide for the executive and for the IT department.

The Opportunity to be an Outcast

Hopefully, you got a job as an IT leader because of management or executive talent – not because you are a great technologist. Unless, of course, you happen to be a great technologist with tons of leadership ability.

Being a technologist is very different from being a leader of people. Technologists have a unique ability to talk to things and make them do what the technologist wants them to do. Leaders are required to communicate with people and work with them to form human teams that efficiently accomplish a goal. That is a huge contrast.

Information Technologists write code, or design hardware or network systems that force electrons and photons to go down specific paths and obey strict rules of operation. And those electrons and photons continue to do what they are told over and over until the system finally breaks or is taken out of service.

Great technologists who are in leadership positions but don't have leadership skills might try to do the same thing with people. I know because I've worked for a few of them who have tried, and failed, to turn me into a robot. The result generally is disaster.

Such an individual in an IT leadership position can be operating at a strong disadvantage in other ways as well. Since IT is such an intrinsic part of almost all enterprises it is essential that an IT leader be able to communicate in business language with the other humans at their peer level across the enterprise. Many of these business people will not have anywhere near the same method of thinking or communication as an IT person. And they shouldn't. So you must be able to communicate with them in their terms to gain their confidence. At the same time you must be able to communicate to your technical team the business needs of the enterprise in a manner that motivates technologists to produce technology that meets business needs. After all, technology without business application is worthless.

An IT manager or executive who cannot create this type of relationship with their peers and promote such a professional culture between their team and the rest of the organization will be a disaster for their team and the enterprise. Eventually, they will be a leadership outcast.

The Opportunity to be a Hero

The information technology executive who can create great internal business credibility has an excellent chance of being invited to meet with external customers. That leader will have proven that they can communicate in a clear businesslike manner the benefits the enterprise can bring to the customer. When this happens, the core leadership team might see that executive as a great asset to use in promoting the enterprise and its products to the end customer. Frankly, if you have a strong business and leadership orientation this is where you can have some of the most fun in your information technology career.

On the other hand, if you are primarily a technologist who would much rather work with things than people you should energetically avoid moving into the management ranks. You will be miserable and make everyone you come in contact with miserable

The excellent manager or executive is focused on relationship management. At the top level this may be one of the most important skills an executive must have. Skill in relationship management begins at the bottom rung of the leadership ladder and may be an especially difficult thing to find in many junior IT managers.

The cultivation of relationships is a critical leadership skill. The more senior the executive the more important relationship management is as the breadth and depth of the executive's influence increases.

At the lower management levels influence is being exercised with the teams being directly managed, with a few peer managers, and hopefully with the next higher management level. This influence spreads across ever widening circles and increases in depth until, at the most senior executive level, it extends into the boardroom, into affiliated enterprises and many layers down into the organization.

If this is a primary skill needed at the most senior leadership level then it is important to cultivate it as a first level IT manager. The most senior executive must be seen as a business person on an equal footing with all in the C-level ranks.

IT Leadership And Career Rules

As the years have progressed I've developed a set of rules for information technology direct contributors and leaders that have come about more through the mistakes that I have made than from my successes. I guess that just helps to confirm what I learned when very young that we learn more from pain than from pleasure. If that is true, then information technology people are among the most knowledgeable techno-masochists in the world.

Things change so rapidly in IT that the professional techie has to learn at warp speed. Then we are expected to forget yesterday's knowledge just as rapidly because it becomes obsolete even faster than the half-life of a boson. Those in IT who have great memories are surely experts at everything that is obsolete!

I sometimes wonder if people get tired of hearing me repeat these rules that, as a group, are focused on technology leaders. However, I do find it interesting to sometimes hear direct contributors, program managers, and even non-geek leaders use these rules when giving guidance.

But, let me get to the reason I am publishing these rules. My over forty years of experience in information technology has shown me that most people in IT leadership positions are great, dedicated, well-meaning technologists who know little to squat about leadership. They are promoted from direct contributor to management because they are so good at managing systems. But very often they don't have a clue about managing teams.

Information Technology is a great career that has become the backbone of almost all enterprises and the world economy. The committed people in information technology deserve inspired, talented leadership. I hope you will find these rules as useful to you as other recovering technology team leaders, and direct contributors, have told me they are to them.

Technologists will often say the devil is in the details. IT leaders have to see this differently. The tempting devil may be the details.

The Rules

- First Rule of Leadership: People will meet your expectations.
- Excellent leaders create excellent organizations.
- Delegate to the limit of your own guilt and other people's pain.
- Keep it simple because it will get complex all by itself.
- Murphy was an optimist.
- One aw shucks (or other word) wipes out a thousand attaboys.
- Technology is easy. Business change is tough.
- Technology without business application has no value.
- Never make a design decision for only one reason.

Those of you who have made the transition from direct contributor to leader know that some days the leadership job is a piece of cake. Other days you will feel totally inadequate to the task.

But, you are the leader and you must lead every minute of every day. Some of you will lead in more critical situations than others. Those who do must make sure your team is confident in you all the time. They will come to trust your decisions and the direction you provide.

There is some magic in this. When your direction is correct your team will succeed. If you are working to be a great leader you will constantly strive to make sure your direction is correct because you won't want to disappoint your team. When your direction begins to take a wrong path your team will still find a path to success because they won't want to disappoint you, the enterprise or themselves.

First Rule of Leadership: People will meet your expectations

The Devil:

I once reported to a leader who worked us like dogs over long hours. He consistently assumed we were going to screw up and was clearly fearful of failure. This tyrant was immediately followed by a true professional who subtly and adroitly provided guidance.

The leader who emotionally flogged us didn't care about our personal life or about family. I suppose he assumed we had no life outside of work because his home life was very visibly unhappy. Most of us who worked for him were frustrated, tense and angry.

This leader was an extremely ambitious technologist who stayed deeply involved in daily system details with little attention to long range goals. He was fully dedicated to making sure every piece of our technology operated at full efficiency. In the process he killed our human efficiency. When he left, we all bought and wore t-shirts irreverently celebrating his departure.

The man who followed this "Captain Queeg" was just the opposite. He was a true gentleman with great command presence who expected the same of the men and women on this staff. He quietly defined his operational expectations and set a mission vision for us to achieve. That example rippled through the organization, as it always does. This executive treated everyone with respect and strove to make sure that we could balance work with our personal lives.

Our efficiency under our new leader improved dramatically. Our technical systems operated better. Morale was high. We became a close, well-oiled machine while he, without seemingly having to put any effort into it, quietly led.

I don't want to over simplify this second leader's great talent or the years that went into him becoming so good at what he did. In the context of this section I simply want to point out that this man had an extremely balanced work ethic and he expected us to have the same. He let go of being a technologist and embraced being a leader. We rose to his expectations just as we fell to the expectations of "Captain Queeg".

The Details:

Every leader at each level in an organization must have an expectation of success, and a vision, for the people or department they manage. That visionary expectation of success must be strongly felt by each team member every day. Each leader, no matter what level they occupy, does not have the luxury of deviating at all from that positive, motivating expectation. They must also be willing to become obsolete at tech detail as they focus on being excellent at leadership.

Those are idealistic words. And I truly believe them and work to practice them all the time.

But welcome to the real world! When was the last time your IT boss told you what their strategic vision was for their team? Your answer might be "How can my boss tell me their vision when they don't even know what vision is!" They might tell you what project they want done today, next week or next month. But that isn't an expectation of success. It is only an expectation of completion.

Many IT leaders come from the techie ranks. As techies we are often project focused. That orientation is frequently carried forward when a person is promoted into leadership. Before being promoted, that techie leader was probably constantly worried, and rightly so, about making small mistakes because those can cause catastrophe in IT and the enterprise being supported.

IT leadership requires expectation reorientation. As a leader you must look for and expect successes, not failures. Anything else will demoralize your team. Each member of your team will already be obsessive-compulsive about avoiding errors anyway. Your leadership expectation must be long term, or vision, focused rather than short term project focused. You must also understand, and accept, that as you progress as a leader you will regress in your knowledge of technical detail.

Unhappily, many IT managers cling to their techie habits rather than embrace leadership habits because they were never trained as leaders. Their expectations may be short term and error prevention focused. Most of us have experienced this much more than we like. Reality deviates from the leadership ideal much too frequently in the world of IT management where most are in management positions because they are super geeks, not inspirational visionary leaders.

The bottom line is this. Excellent leaders understand that people will meet the expectations of their leader, whether they are high or low!

Excellent leaders create excellent organizations

The Devil:

I once worked for an enterprise with a headquarters function of about six hundred people and ten very important remote offices with five to thirty highly skilled people at each location. To make a long story short, our organization had just received an extremely bad business rating and our chief executive, who happened to be my direct boss, was fired.

A new chief executive was brought in from outside. I was the most junior member on his staff, but my team had received the highest rating from the reviewing group. Much to my dismay the new chief executive appointed me to the unpleasant task of forming a small team to do our own internal review of the entire organization and recommend strong actions to quickly improve our performance. This was a full-time, six-month assignment that put my team and me on the road to every remote office.

During those six months I learned the most important lesson in my career about leadership! As we traveled around I found out that the locations that were top performers had the best leaders. Without exception, the performance of the organization directly reflected the performance of the top executive. If the top executive was mediocre or poor, the performance of the organization reflected that of the top executive even if the rest of the people were excellent and self-motivated. These top leaders provided, and were the only ones who could provide, vision, the proper culture, and the team motivation and desire to create an Infrastructure to achieve that Vision.

This was a startling lesson that I frankly had trouble absorbing. Once I finally understood and accepted this lesson, I was able to report to my very seasoned boss a clear resolution to our organizational problem. Naturally, he already understood this. He just needed good data on where and who the problems were. Some key leadership replacements were made, and the next report from the reviewing team was the best the organization had ever received.

The Details:

How many IT managers practice the following bullet points, or even know they exist?

- Excellent leaders understand that people will meet the expectations of their leader – low or high.
- The best leaders provide a vision that excites the team to action and gives them clear objectives.
- Excellent leaders minimally manage and they don't administer at all!
- Leaders know that if they assume a management role rather than an executive role then the people who work for them will expect to be managed, and they will act as if they should be managed.
- Leaders know that if they unnecessarily take on administrative tasks they won't be in the proper frame of mind to lead.
- Excellent leaders stay in their leadership role all the time!

It isn't their fault if IT managers don't understand these things. They are seldom, if ever, taught them. Nor do they have much opportunity to learn by example. Almost all are promoted to management positions because they are miracle working technologists who talk to technology devices much better than they talk to people.

So, here is a bit of heresy. Technology managers must be much better leaders than they are technologists!

In fact, senior IT leaders don't have to be technologists at all. The best leaders can be equally adept as the chief executive for the space shuttle program or being in charge of the interstate highway system.

The message for IT leaders, and the people who select them, is that excellent leaders create excellent organizations.

If you are in a position to hire a new leader at any level look at who they worked for in the past. Great leadership is learned by working for great leaders. Harold S. Geneen who was president of International Telephone and Telegraph put it well when he said, "Leadership cannot really be taught. It can only be learned."

A great and well-meaning technologist can create chaos when placed in IT leadership positions if they have weak leadership skills. On the other hand, excellent leaders who know nothing about technology can establish amazingly efficient IT teams that produce superior products.

Delegate to the limit of your guilt and other people's pain

The Devil:

An organizational transfer put me in a new assignment with huge management and planning responsibilities at a young age. The team I was leading was a bit larger than my previous one, overloaded, and very good at what they did. I had such a diversity of responsibilities that I hardly knew where to begin. Instead of being a team manager and excellent senior staff member I dove into being a prodigious worker. I became a member of my team instead of the leader of my team.

Happily, one of the highly experienced people working with me on the executive staff, and very senior in years and experience, saw the error of my ways. From him, I learned the value and art of delegation and the necessity to step away from work in order to be an effective manager. This was a difficult lesson for me to absorb, but when I finally did, the team operated more efficiently. I became a better executive staff member making a contribution to organizational efficiency and evolution.

The Details:

Delegation may the most difficult thing for new leaders to learn and for more experienced leaders to persist in. There can be a lot of guilt associated with delegation. An empathetic manager will feel compelled to jump in and take on part of the workload of their subordinates if they are overloaded. This is understandable, and it is a damaging mistake.

Your task is to use your empathy in a better manner and delegate to the limit of your own guilt and other people's pain. This will provide you with the space to employ your empathy to force yourself to find solutions so you can eventually relieve your guilt and their pain.

A leader must be in the leadership role all the time whether they are a first level manager or the CEO. A person in a management role might be tempted to take on non-leadership projects. But the minute a leader does that they lose their focus on where they are taking the team. Either lead or follow! Don't try to do both.

The problem, which seems to be especially prevalent in IT, is that a leader who might have the skills to do a technical job better than one of their team members might just dive into it. Or, they might feel that the team is overloaded so they take on a direct contributor task to share the burden.

Another challenge to leaders who were previously direct contributor technologists is the fear of becoming technologically obsolete. Frankly, IT leaders have a choice. Become great leaders and weaker technologists or continue to be great technologists and inadequate leaders. No one can do both!

Clinging to direct contributor habits and knowledge of technical details are horrible and alarmingly frequent IT leadership errors.

Constantly delegating is tough to do, especially for junior IT managers who might feel great empathy for their team's situation. But diving into technical details inhibits your ability as a leader to properly assess the situation so you can find better ways to get the job done. You will do more harm than good.

Don't lose your professionalism and get exasperated by tough situations. Never give up a calm, commanding leadership role. If you want to practice leadership through exasperation then try time travel to the Middle Ages and persecute villagers. In the long run you will be persecuting your employees when you focus on work solutions instead of leadership solutions.

Information technology managers may find it especially difficult to delegate because they are taught to be hands-on engineers and they love technology challenges. Please be assured that management challenges are much tougher and more rewarding than technology challenges. And if you handle your management role properly you will multiply your technology contribution by creating a productive, happy environment for your team.

Follow the advice of Ronald Regan who said, "Surround yourself with the best people you can find, delegate authority, and don't interfere."

Keep it simple because it will get complex all by itself

The Devil:

I once worked with an excellent vendor corporation that was a great business partner for our enterprise. This vendor was a large telecommunications organization that provided us with consistently good advice and products. Then we got involved with a product team in this enterprise that was developing an advanced system that promised to reduce our cost of operation and simplify our technology environment. The problem we ran into was that the product manager kept adding new bells and whistles which complicated the system more and more as time progressed.

In its simple form the product would have been great. It finally collapsed under the weight of its own complexity. The vendor lost a great sales opportunity to a number of customers. We fell behind in our technology deliverable and had to take an entirely different, and much simpler, systems direction.

It was worse than a train wreck. It was a plane wreck.

The Details:

The corporate life expectancy of CIOs is horrible. In many cases they lose their jobs because they engage their enterprises in technical solutions that promise the world but dramatically under deliver because they are so complex and consequently filled with technical holes.

In general, this type of thing is a trap for any executive. But it is especially dangerous for IT executives because the career field is already so challenging in its complexity and evolutionary pace.

The message for IT executives at all levels is to keep your technical environment as simple as you can because it will get complex all by itself. A simple, straightforward solution won't be as full of career limiting hazards as one that starts out as vastly more complex.

IT work is already complex by its very nature. After all, the job of your technical wizards is to precisely direct the activities of subatomic particles along pathways that can't be seen to control equipment you might never touch at locations you probably won't visit. In another age this would be called witchcraft!

Don't make this stuff more complex than it has to be when you are working to use it to support a business initiative. You will bury yourself, your tech team and your enterprise and get yourself fired for enthusiastic, well-meaning incompetence.

It is easy for IT managers, especially first level managers, to get sucked into approving complex technical solutions proposed by their technical team members and by vendors. After all, complex solutions are fun for technologists. They also provide lots of long-term job security for workers because they have to be tweaked and maintained and fixed.

But just the opposite is true for the technology manager who works with the team to implement these complex solutions instead of leading them to the reduction of complexity. Complex solutions will always become even more complex because that is the way technology, and everything else, is.

They will almost inevitably lead to a short life expectancy for information technology managers and executives.

You can be like Achilles and lead a short exciting life as an information technology leader. Or you can be the mountaintop sage who people come to for many years for wise advice on how to use technology to make your enterprise great. That long life as a technology leader will still be filled with its exciting moments.

To quote General Colin Powell, "Great leaders are almost always great simplifiers, who can cut through argument, debate, and doubt to offer a solution everybody can understand."

Murphy was an optimist

The Devil:

After taking great precautions I was leading my team in a major weekend network cutover. It involved two cross country telecommunications trunks that carried over fifty percent of the revenue for our organization. Our planning was meticulous leaving nothing to chance. Or so we thought. We had set up some temporary routes on completely diverse paths to maintain traffic while we made our transition.

Then two things happened almost simultaneously that would make any technologist fully believe that Murphy was an optimist.

One of the temporary trunks failed. The good news was that it was dynamically rerouted to a new path with only seconds of interruption. The bad news was that the two diverse paths now met at just one point along a route over one thousand miles long.

Then a tanker truck driver with a full load of gasoline slid off the road and crashed into that one point where the circuits met. Over half the revenue for our enterprise stopped flowing for the next twenty-four hours until the new routes were in place.

I still believe that Murphy had to be driving that tanker truck.

The Details:

Almost every information technology professional knows about Murphy's Laws. Certainly it doesn't take but a very short time (measured in millidays) for an information technologist to come to the conclusion that bad things happen in the information technology profession far more frequently than even the pessimistic expectations of Murphy, whoever he really was.

The challenge for IT leaders at every level is to keep a steady head and an air of totally professional calmness during calamity. While you will do everything to avoid it, you simply have to understand that your enterprise's financial IT systems will only fail when they are needed the most.

If you have created the proper work environment in your organization you will not have to frantically wave your arms at your IT gurus to get them to fix the problem. Your dedicated techies will be quietly working on their own at warp speed while sweat drips onto their crumb-filled keyboards. Your job is to protect them from your peers and superiors in the enterprise who will be making meaningful, and possibly impolite, gestures your direction.

All of this makes almost all IT professionals superstitious. You, as their leader, need to understand that your employees will desperately search for the nearest wood to knock on if you complement them on how well their systems are running. It will be helpful, and courteous of you, if you at least keep a block of authentic wood handy for emergencies.

You, as the radically calm information technology leader, must simply accept the fact that you are in the most thankless of professions. You and your team will receive little gratitude for the miracles they perform every day. Disaster will inevitably strike at the worst possible time. You will eventually understand that while most see Murphy as a pessimist, he had never peered down the dark tunnel of endless terror inhabited by today's information technology professionals.

As an information technology leader it is your job to create a positive, productive calmness in this intrinsically chaotic environment. You can't just keep the block of wood handy for the superstitious to knock on. Your people have to know that you are dedicated to keeping their work environment safe as you lead them to success. With your help they will stay productive and dedicated in spite of the innate stressful characteristics of information technology.

As General Bruce Clarke said, "Rank is given you to enable you to better serve those above and below you. It is not given for you to practice your idiosyncrasies."

Challenge the team in a positive manner

The Devil:

I once worked directly for a middle level manager we were all thrilled to see leave the organization because he had created such serious problems. Our new manager immediately took several actions to correct the issues. One thing he did came to impress me even more as the years went on than it did at the time. He made the assumption that we were all expert professionals who wanted to be the best at what we do. He made it clear that he had high expectations of all of us, that he had great respect for us, and that he knew we could succeed. We did succeed beyond even our own expectations which meeting his.

I have also seen a new chief executive come to an organization with an erroneous opinion, provided by others, that he was taking over a group that had a dysfunctional team with some individuals who needed to leave. We had a miserable two years during which we all struggled to perform well under an umbrella of suspicion. Finally, he reformulated his opinion of us and began to understand that his expectations could be much higher. We never did great under this man, but by the time he left we were doing well. We met his somewhat limited expectations.

The Details:

The excellent IT leader will challenge the team in a positive manner.

Information technology is complex and fast moving. No member of the team is going to fully keep up with the latest technologies and the best way of doing things. Technologists tend to love what they do because it is so challenging. They also want to do everything right because they almost universally have such pride in their profession and what they provide for their enterprise. They are probably more critical of themselves than they should be and are generally striving for more perfection than is economical.

Everyone at all levels in information technology knows the criticalness of their work and seldom has to be reminded. If a computer crashes, a production line might stop and revenue may stop flowing. A software glitch can cause a massive selloff in stocks, or trains and planes to collide with each other resulting in deaths for many people.

IT is a stressful career field and it must be led by managers and executives who can motivate in a positive, respectful and logical manner. The superior IT leader will understand this.

Coach John Wooden observed that, "When you punish your people for making a mistake or falling short of a goal, you create an environment of extreme caution, even fearfulness. In sports it's similar to playing 'not to lose' - a formula that often brings on defeat."

Some executives think they are only doing their job properly if they are constantly looking for failures rather than successes.

That seems to be a special problem with IT managers and executives.

The problem is complicated by the fact that information technology systems play such a major role in the infrastructure of a large percentage of enterprises. Anyone who has been in IT for just a few short years knows that one aw shucks wipes out a thousand attaboys. Consequently, many IT managers and executives have an overwhelming temptation to constantly be personally looking for technical errors before they become business impacting issues.

That negative form of leadership makes people nervous and distrustful of the boss. That nervousness slows work down as employees start second guessing themselves or become reluctant to ask for help when they run into problems. Even worse, they may not tell their manager when a problem that is debilitating to the enterprise is occurring until it is out of control.

General Dwight D. Eisenhower said, "You do not lead by hitting people over the head – that's assault, not leadership."

Technology is easy. Business change is tough

The Devil:

A key opportunity that one of my past CIOs once brought to the team was something he called the "blow up the warehouse" initiative. Margins were narrow in the business we supported and costs needed to be shaved if we were going to continue to be competitive or perhaps even survive. The new idea was to move product directly from the manufacturing site to the customer without having to pay for extended storage in large warehouses. The CIO understood the potential deliverables that the IT team could provide. He also fully understood the core enterprise.

The direct result of this effort he initiated was significant reduction in storage costs and improvement in tracking products between the plant and the customer. One indirect byproduct was provision of additional information to inside and outside sales so that they could provide our customer with more accurate information on delivery dates. A second indirect benefit was the new ability for our outside sales people to immediately provide this information personally to the customer at the time of the sale. The strong competitive and financial benefit was an increase in sales because we provided more timely and more accurate information to the customer than our competition could.

We partially "blew up the warehouse" and we did it in conjunction with layering new applications on top of a strong IT infrastructure that was well-integrated with the business infrastructure. We wouldn't have done any of this without a CIO who was intimately knowledgeable of the core enterprise and who knew the capabilities and limitations of their team.

The Details:

Congratulations to you if you are an information technology manager and are still reading this after seeing the potentially off-putting headline of this section that says technology is easy, business change is tough.

Okay, maybe that should be a comparative more than an absolute headline. But it got your attention, didn't it?

The comparative statement would be "business change is much tougher than managing technology".

The point is this. The decisions that technology leaders make must put the business ahead of technology. To that end it is important to choose leaders at all levels of the IT hierarchy who will almost automatically make decisions with that priority in mind. A person in a leadership position who focuses more on the technology than on the influence of that technology on the business is going to weaken the entire enterprise. Even the first level manager of a small technology team will have some effect in this regard, positive or negative, on the entire organization.

Information technology teams install voice, wireless, network, data and computing systems and provide them as tools to help people get their daily tasks done. These are challenging devices to install properly and keep working reliably. However, the technologist who is fully integrated into the business of the enterprise will soon realize that the real challenge is using the technology to bring marketplace value through business change.

The CIO and team who bring the full power of technology value to the business will dramatically improve the enterprise's position in their industry and the position of the IT team in their enterprise.

Now that the heresy is out of the way, it is important to make it clear that technology leaders must be chosen who can integrate the IT team into the core business planning process. That leader must mesh IT vision with business vision and IT infrastructure with business infrastructure. Along the way, business vision and infrastructure are enhanced by the CIO's team if they bring greater value to the business than would have been realized without a strong IT leader who was primarily a strong business leader.

The "old AT&T" used to have a motto: "The System is the Solution". That is pretty much the point here. The CIO team's technology solutions merge with the business solutions into a single system for the enterprise.

Technology without business application has no value

The Devil:

I'm not a software guy. But I was once assigned to lead a software team for a short time. I thought my boss had lost his mind when he asked me to take on this role and solve the problem.

The software team had been assigned to build an information analysis system to assist senior executives in decision making. I was only involved with the project for a few hours when I discovered just how important this system was. By this time I knew my boss had lost his mind.

The problem was that the system was useless to the executives. They didn't understand the output or how to get to the information that was needed.

The software gurus didn't even begin to understand the problem. From their perspective all the information was available and the system worked perfectly.

Within a day I understood why my boss had put me in charge of the project. Only a geek could have deciphered how the software techies were presenting the information. With just a few tweaks the output was reformulated into a method the executives could understand and easily use.

A great technology system that had been valueless suddenly became indispensable when made into a proper business tool.

The Details:

Technology is one of the most rapidly evolving components of our society. Information technology may lead the pack in speed of change. The IT executive is going to have to wisely choose the person on their team who will be charged with predicting the IT strategic target that most optimally supports the business strategic vision.

The strongly business-oriented IT executive will be primarily focused on knowing where the core enterprise is going over the next few years. This will have to be discovered in part through the strong relationships the IT leader will necessarily build with all members of the executive team. Once long term clarity of the business strategic target is achieved the IT chief executive can take that information back to their team with the charge of determining how to use IT to enhance the enterprise's ability to reach, and perhaps exceed, the enterprise strategic target.

The last sentence of the previous paragraph is the tough part for the IT leader. The business path will have been primarily laid out by the core enterprise executives. Estimating how the most rapidly evolving of technologies can be used to assure enterprise success is extremely difficult.

Many members of the IT team will have a tendency to look at technology as its own end. There may only be a few, and the chief IT executive has to be one of them, who will understand that technology is just a tool used to guarantee that the enterprise core business will be much more successful than that of its competitors. The best IT executives, and planning teams, will find ways to use technology to enhance the business strategic target and sell those ideas to the core team.

The business life expectancy of CIOs is relatively short compared to their enterprise peers. In many cases this is because the CIO isn't fully integrated into the core enterprise and doesn't make it clear to their IT team that technology without business application has no value! When speaking of technology to their business peers the CIO must make it known that they and their team are business people first and technologists second.

At the same time that the IT executive is leading their team in the development of a business supporting strategic target, an evolutionary path to that target has to be put in place. The chief executive who assures that their team does this well will keep costs down as they avoid technological dead ends. They will avoid developing and implementing interesting and fun tools that don't necessarily have application to the enterprise.

Rear Admiral Grace Murray Hopper observed that "You manage things; you lead people."

Just a note of interest. Admiral Hopper is credited with popularizing the term "debugging" for fixing computer glitches. This came about in 1947. Her team found and removed an actual moth from a computer that was malfunctioning. You might be interested in reading about this woman who was that rare combination of a great leader and accomplished technologist.

Never make a design decision for only one reason

The Devil:

Not too long ago, in relative terms, I worked for an IT chief executive who had a million ideas about what he wanted to do. We were all thrilled that he came to us with such enthusiasm. As he shared his ideas with us it became clear he was focused clearly on the applications we could deliver to our customers. That was good news as far as it went.

The challenge that we all encountered was that he had little concept of the need for an evolving underlying infrastructure and strategic vision needed as a platform for his ideas. As a good team should, we showed him how the ideas he had would be easier to implement and be delivered more efficiently if we first worked on an infrastructure that would support his ideas. It became clear fairly quickly that he had no concept of the difference between infrastructure and application. He had always been very application oriented and the infrastructure had been invisible to him. Our attempts to educate him on how a strong infrastructure could help him reach his application goals fell on deaf ears.

Frankly, I was stunned that he had been promoted to the IT executive ranks without having the essential knowledge of infrastructure. None of us should have been too surprised because he worked with a pretty narrow, short range focus, and his appointment had been more political than professionally earned. I'm sure many of you reading this book may have encountered the same problem.

He finally left having accomplished nothing except to produce a tremendous amount of useless turmoil.

The Details:

If you are an expert technologist you will make sure that your designs and solutions are extensible and integrate well with other technology components. You will be looking to the left and right of your design to see how everything else in the system functions so that you can enhance the entire system and avoid doing things that detract.

For the information technology manager another dimension is added to this design requirement. Enterprises are complex. They, in many logical, political and emotional ways are even more complex than the information technology that supports them.

The information technologist who is promoted to a management position will already know that IT is complex. However, it seems that much too often these new managers believe that management and leadership are easy compared to technology work.

The consciously competent IT manager will quickly learn that the application of technology to broad enterprise goals is almost overwhelming in its complexity. As a techie they were focused on their piece of the requirement. As a leader they are exposed to intertwined business requirements. While some business elements may be satisfied by a particular tech solution others may be weakened. The new tech manager will have moved from the platform of the two dimensional puzzle into the world of the Rubik's cube.

An excellent technology developer will always keep a business system solution rather than a component solution in mind when completing a design. The IT manager will be in an even more complex world. They will be guiding their team in a wide ranging set of extensible system solutions that work in the technical world with great application in the enterprise universe.

The IT leader who can perform effectively in this "Rubik's cube" environment will do a great service to the enterprise they are supporting as well as to the tech team they are leading. The enterprise will receive the benefit of a tech leader who understands how to apply technology to help the enterprise potentially exceed its mission and revenue goals. The technical team they lead will receive recognition from their internal customers for the excellent service they are providing to the enterprise.

As an information technology leader you must understand that the technology team's best design solution is the one that best helps the overall enterprise reach its mission goals. As a leader you must help your team find that solution, or integrated set of solutions, then get out of their way while they implement it and feel great about themselves.

Lao Tzu knew that "A leader is best when people barely know he exists, when his work is done, his aim fulfilled, they will say: we did it ourselves."

Wrapping it up

Information technology is an incredibly complex, stressful career field filled with phenomenally dedicated people. It is good that these workers are so professional because IT is also becoming the often invisible backbone of corporate, private, non-profit and government enterprises across almost the entire planet. When IT fails, then some part, or all, of the enterprise supported by that IT element comes to a screeching stop.

IT workers, the enterprises they support, the customers and the population in general deserve excellent IT leadership. My concern is that the leadership element of information technology is its most vulnerable part. This is because the people placed in IT management and executive positions are often put there because they are great technologists rather than great leaders. I have more than once seen excellent leaders passed over for middle and senior IT management positions because "they weren't technical enough".

I'm not so naïve as to think that this little book will change all this. But I hope it will influence management selections to some extent. Every enterprise and their customers deserve information technology leaders who will provide tools that will make the enterprise great and perhaps have a positive effect on profitability, end product quality, customer satisfaction and the economy in general.

Yes, I am a bit of an idealist!

I know I'm right about all this because I have been in responsible leadership positions in information technology organizations for over forty years. During that time I have worked with or for a number of leaders. A few were IT experts and also excellent leaders.

But much more often the best IT executives were not technologists at all. They were hired into the technology group because they were excellent leaders. When that happened everything worked much better. Some were non-technologist military officers. The best I ever worked for was an attorney who barely understood technology at all. But could he lead!!! And the team he led became great!

The worst was an MBA. Go figure!

Information technology is a great and demanding career filled with amazingly dedicated people. These professional technologists deserve professional leadership.

General John J. Pershing knew that "A competent leader can get efficient service from poor troops, while on the contrary an incapable leader can demoralize the best of troops."

One Last Devil:

In my very first information technology management assignment I was working for a well-seasoned executive who happened to inherit some inexperienced, talented and aggressive young leaders. There was one discussion we were involved in that became professionally contentious with strong good opinions on both sides. We couldn't reach resolution so we did the right thing and went to see the boss. Each side stated their case as he listened carefully and asked several questions. We could see that he understood the issue and the reasons for our opinions. Eventually, he said something like, "OK, I understand. And here is what we are going to do."

After he delivered his decision the discussion immediately resumed and the boss let it continue for thirty seconds or so. I remember so clearly that he held up his hand for silence and then quietly instructed us. "Gentlemen, this has been a good discussion. What you have failed to understand is that I have made a decision. That means the discussion is over. You are all excused and you may now follow the direction I have given."

What a great lesson gently and effectively given. I have carried that lesson throughout my entire career. I suspect that it originated with a wise leader thousands of years earlier and was passed on over time from person to person.

www.ingramcontent.com/pod-product-compliance
Lightning Source LLC
Chambersburg PA
CBHW071643170526
45166CB00003B/1417